Power PRACTICE™

Parts of Speech and Punctuation

5–6

Written by

Pamela Amick Klawitter, Ed.D.

Editor: Pam VanBlaricum
Illustrator: Jenny Campbell
Production: Acorn Studio Books
Cover Designer: Barbara Peterson
Art Director: Moonhee Pak
Project Director: Linda Schwartz

Table of Contents

Introduction

Each book in the *Power Practice*™ series contains dozens of ready-to-use activity pages to provide students with skill practice. Use the fun activities to supplement and enhance what you are already teaching in your classroom. Give an activity page to students as independent class work, or send the pages home as homework to reinforce skills taught in class. An answer key is provided for quick reference.

Parts of Speech and Punctuation is filled with activities designed to reinforce the parts of speech and capitalization and punctuation rules for grades 5 and 6. Each page introduces the rules for a specific skill, followed by reinforcement activities. The book covers the following standards-based skills:

Sentence Structure

Recognizing and writing declarative, interrogative, imperative, and exclamatory sentences.

Identification and Correct Usage of Parts of Speech

Recognizing and correctly using nouns, verbs, pronouns, adjectives, adverbs, prepositions, conjunctions, and interjections.

Capitalization Rules

Rules for capitalizing proper nouns and adjectives, abbreviations and initials that stand for proper nouns, and titles.

Correct Usage of Punctuation

Understanding the correct usage of commas, apostrophes, quotation marks, colons, semicolons, and hyphens.

Review Activities That Put It All Together

Use these activities and watch your students' writing skills and standardized test scores improve.

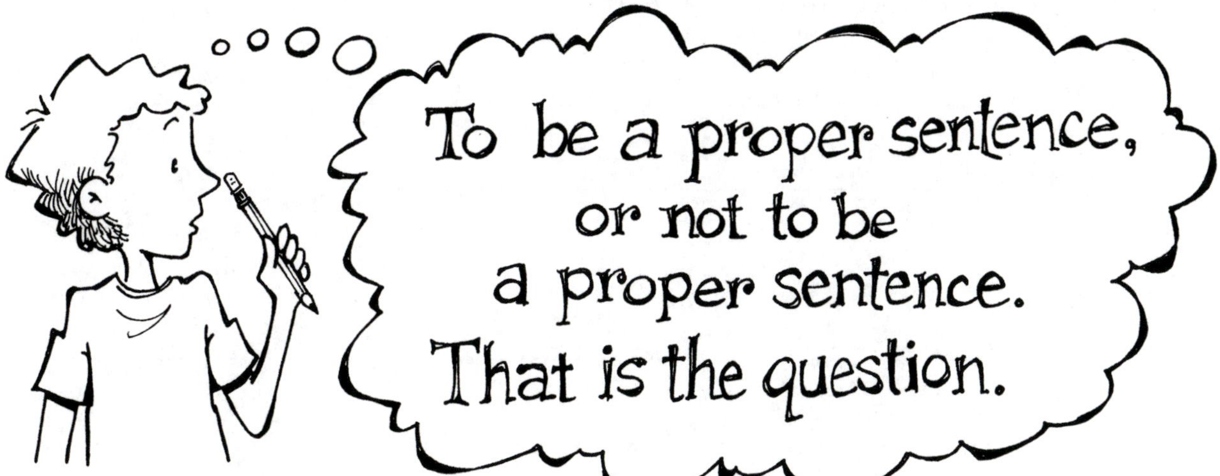

Name ______________________________ Date ______________

Sentence or Not?

RULE

A *sentence* is a group of words that expresses a complete thought.
A *fragment* is a group of words that does not express a complete thought.

Circle **S** for each group of words that is a *sentence* or **F** for each group of words that is a *fragment*.

S F 1 His best friend just returned from an Alaskan cruise.

S F 2 On the edge of her desk.

S F 3 The day care center next to the park.

S F 4 The teacher's desk is in the front of the classroom.

S F 5 Most of the girls on the team.

S F 6 The local press reported the story three weeks ago.

S F 7 Three of the major league teams.

S F 8 Found it hard to believe.

S F 9 Ate all the cupcakes before the party.

S F 10 The puppy gnawed a hole in my new shoe.

S F 11 From five different countries.

S F 12 During the hottest part of the day.

Rewrite five of the fragments to make complete sentences.

1 ______________________________

2 ______________________________

3 ______________________________

4 ______________________________

5 ______________________________

Name ______________________________ Date ______________

Sentence Sense

RULE

There are four basic types of sentences:

- A *declarative sentence* makes a statement or tells something. It ends with a period **(.)**.
- An *interrogative sentence* asks a question. It ends with a question mark **(?)**.
- An *imperative sentence* gives a command. It ends with either a period **(.)** or an exclamation point **(!)**.
- An *exclamatory sentence* expresses a strong feeling. It ends with an exclamation point **(!)**.

Place the correct end mark after each sentence. Label each to tell which type it is—declarative, interrogative, imperative, or exclamatory.

1. The shape and size of a bird's beak depends on its diet___ ______________
2. Believe it or not, an Arctic tern migrates more than 22,000 miles every year from the Arctic all the way to the Antarctic and back again___ ______________
3. Is it true that a wandering albatross can live 80 years___ ______________
4. Help rescue birds that are affected by oil spills___ ______________
5. Why do birds have an extra transparent eyelid___ ______________
6. Do kiwis really live in underground burrows___ ______________
7. Webbed feet help water birds propel through the water___ ______________
8. Don't buy rare and endangered birds to keep as pets___ ______________
9. Wow, we clocked that ostrich at sixty miles per hour___ ______________
10. Did you know the extinct moa grew to be over ten feet tall___ ______________
11. A flamingo is pink because of chemicals in the food it eats___ ______________
12. Never damage a wild bird's habitat___ ______________

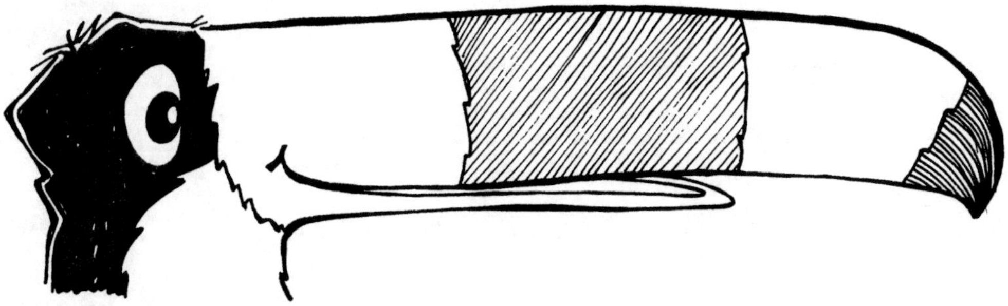

Name ______________________________ Date ______________

Write It Down!

Practice: Remember the four basic types of sentences: *declarative, interrogative, imperative,* and *exclamatory*. Choose three topics from the topic box. Write one of each type of sentence about each of the three topics.

Topic Box

best friend	favorite pet	siblings	books
my class	hobbies	foods I hate	sports

Topic 1: ______________________

declarative: ______________________

interrogative: ______________________

imperative: ______________________

exclamatory: ______________________

Topic 2: ______________________

declarative: ______________________

interrogative: ______________________

imperative: ______________________

exclamatory: ______________________

Topic 3: ______________________

declarative: ______________________

interrogative: ______________________

imperative: ______________________

exclamatory: ______________________

Name ____ Date ____________________

The Subject Is

RULE

- The subject of a sentence tells the person, place, thing, or idea that the sentence is about.
- The complete subject includes all the words that identify who or what the sentence is about.
- The *simple subject* is the main noun or pronoun in the subject.

Example: The highly venomous Gaboon viper lives in Africa.
(simple subject: viper; complete subject: The highly venomous Gaboon viper)

Underline the *complete subject* in each sentence below. Write the *simple subject* on the line following the sentence.

1. The timber rattlesnake uses its rattles to scare off intruders. ____________
2. Meat-eating animals use special teeth to pierce, slice, and bite. ____________
3. The fluffy young owl has down to help trap heat next to its body. ____________
4. Flat tropical grasslands are located in central Africa. ____________
5. The unusual coconut crab can climb a tree to get its food. ____________
6. The brand new Metro Zoo just opened six interactive exhibits. ____________
7. A fully grown Florida manatee weighs about a ton. ____________
8. The high, misty mountains of central China are home to pandas. ____________
9. One of the scariest nighttime sounds is the low hoot of an owl. ____________
10. The tallest water bird on the planet is the flamingo. ____________
11. The white-tailed deer fawn has a spotted coat for camouflage. ____________
12. Huge herds of grazing buffalo once covered the Great Plains. ____________

Name ______________________________ Date ______________

Predicate Predicament

RULE

The *predicate* of a sentence tells what the subject is or does.

- The *complete predicate* includes all the words that tell what the subject is doing or that tell something about the subject.
- The *simple predicate* is the verb in the sentence.

Example: The highly venomous Gaboon viper lives in Africa. (simple predicate: lives; complete predicate: lives in Africa.)

Use each *complete predicate* from the box in a well-written sentence. Circle the *simple predicate* in each sentence.

1. ______________________________
2. ______________________________
3. ______________________________
4. ______________________________
5. ______________________________
6. ______________________________
7. ______________________________
8. ______________________________
9. ______________________________
10. ______________________________

Complete Predicates	
1 plays softball with the Marauders	6 resembled a small helicopter
2 was the size of a grapefruit	7 built houses from adobe bricks
3 writes for the local newspaper	8 looked pale blue in the sunset
4 fell off the top shelf	9 chewed its way out of the box
5 ate six slices of pizza	10 fit in the palm of my hand

Name ______________________________ Date ______________

What a Combination

RULE

Using different types of sentences adds interest and variety to your writing.

- A *simple sentence* has one subject and one verb.

  Example: A lizard slithered across the path.

- A *compound sentence* is made up of two or more simple sentences with related ideas. They are usually joined by a comma and connected by a conjunction such as *or*, *and*, or *but*.

  Example: A blue whale is the largest marine mammal, but an elephant is the largest land mammal.

Combine each pair of simple sentences into a compound sentence. Circle the conjunction in each of your sentences.

1. A coyote is a member of the dog family. It is less than half the size of a wolf.

   ____

2. An eel looks like a snake. It is really a type of fish.

   ____

3. An Asian elephant is ten feet tall. An African elephant is even bigger.

   ____

4. A hermit crab is a crustacean. It has five pairs of jointed legs.

   ____

5. The female hyena is bigger than the male. She leads a group called a clan.

   ____

6. Seals are quite clumsy on land. They are fast and graceful in the water.

   ____

7. A male ostrich stands eight feet tall. It can weigh more than 400 pounds.

   ____

8. Grasshoppers have wings and can fly. They usually walk and hop.

   ____

9. Dolphins are toothed whales. There are over thirty different species.

   ____

10. Arabian camels have one hump. Asian camels have two humps.

   ____

Name ______________________________ Date ______________

Identifying Nouns

RULE

A *noun* names a person, place, thing, or idea.

- A *common noun* names *any* person, place, thing, or idea.
- A *proper noun* names a *specific* person, place, thing, or idea and always begins with a capital letter.

Fill in each blank with the correct type of noun.

1. ____________ (proper) played Robin Hood in this year's school play.
2. Serena and her friends looked for ____________ (common) on the beach.
3. Three men from ____________ (proper) tried to swim across the lake.
4. The ____________ (common) traveled to ____________ (proper) last year.
5. ____________ (proper) hiked completely around ____________ (proper).

Write a sentence using each pair of nouns below. Use the *common noun* first in each sentence.

1. Mars, telescope

____

____

2. llama, Peru

____

____

3. Chicago, buildings

____

____

4. cabin, Bill

____

____

5. mountains, Colorado

____

____

Name ______________________________ Date ______________

How Many?

RULE

A *singular noun* names one person, place, thing, or idea.

A *plural noun* names more than one person, place, thing, or idea.

Remember that a *common noun* names *any* person, place, thing, or idea. A *proper* noun is specific and always capitalized.

Circle all the nouns in each line that fit the description.

1 Plural:	wharves	radio	banjos	tattoo	spies	rodeos
2 Singular:	hat	boxes	hero	studio	echo	piano
3 Singular:	mosquito	burgers	salad	jobs	ballerina	mice
4 Plural:	moss	waltz	phrase	perches	ladies	flies

Review: Choose four words from the word box that fit each description.

1. **Plural and common:** ____________, ____________, ____________, ____________
2. **Singular and proper:** ____________, ____________, ____________, ____________
3. **Plural and proper:** ____________, ____________, ____________, ____________
4. **Singular and common:** ____________, ____________, ____________, ____________

Word Box

pencil	Garfield	geese	Seattle	equator
Darla	cities	Canadians	Buddhists	leaves
Neptune	Toyotas	kangaroo	Brazil	Yankees
women	movies	Mondays	seamstress	frontier

Name ______________________ Date __________

More Than One

RULE

A *plural noun* names more than one person, place, thing, or idea. To make most plurals, simply add an *s* to the singular (dog → dogs). There are some exceptions:

- If the noun ends in *s*, *ch*, *sh*, *x*, or *z*, add *es* (witch → witches).
- If the noun ends in a consonant + *y*, change the *y* to *i* and add *es* (country → countries).
- If the noun ends in a vowel + *y*, simply add *s* (boy → boys).
- If the noun ends in *f* or *fe*, sometimes you simply add *s* (chief → chiefs; giraffe → giraffes), and sometimes you change the *f* to *v* and add *es* (leaf → leaves; knife → knives).
- Some nouns have the same form for singular and plural (*examples:* sheep, deer, fish, scissors).
- Some nouns are irregular (*examples:* tooth → teeth, goose → geese, man → men, child → children, mouse → mice).

Write the plural form of each noun below. You may use a dictionary.

1. wolf ______________
2. woman ______________
3. shelf ______________
4. guy ______________
5. foot ______________
6. elf ______________
7. family ______________
8. monkey ______________
9. journey ______________
10. match ______________
11. fox ______________
12. church ______________

Write the singular form of each noun below. You may use a dictionary.

1. oxen ______________
2. cacti ______________
3. loaves ______________
4. berries ______________
5. tomatoes ______________
6. calves ______________
7. lives ______________
8. moose ______________
9. salmon ______________
10. flies ______________
11. fairies ______________
12. dice ______________

Name ______________________ Date ____________

Lights, Camera, Action!

RULE An *action verb* expresses action—tells what is being done by the subject. To find an *action verb*, find the subject and ask, "What did the subject do?"

Example: Trina **rode** her bike to the store.
(The subject is *Trina*. What did she do?)

An *action verb* usually shows *physical* action (ran, ate, built), but some show *mental* action (thought, believed, wished). A few action verbs show possession or ownership (have, own).

Underline the action verb in each sentence. On the blank following each sentence, write another action verb that changes the sentence's meaning.

1. Jared gathered his gear for the camping trip. ____________
2. The soccer team practiced for the upcoming tournament. ____________
3. The bus driver parked at the corner of Fifth and Main. ____________
4. My teacher assigned five questions yesterday. ____________
5. Several families of ducks waded near the lake shore. ____________
6. The giraffe munched the leaves on the highest branch. ____________
7. The hikers found a shelter beneath the trees. ____________
8. Sarah read the French menu to her mother. ____________
9. Don rebuilt his house after the hurricane. ____________
10. The toddler broke his new toy fire truck. ____________
11. Aunt Joyce decorated all the cookies before the party. ____________
12. John washed the window in the family room. ____________

Name ______________________________ Date ______________

Help!

RULE

Sometimes the verb in a sentence is not just one word. The main verb can be written with *helping verbs* to create a verb phrase. These are common helping verbs:

- *be* → am, is, are, was, were, be, being, been
- *have* → has, have, had, having
- *do* → do, does, did, done, doing

These are other helping verbs: *may, might, must, can, could, shall, should, will, would.*

Examples: I *am flying* to Miami for a two-week vacation.
He *has been singing* that song all day long!

Underline the verb phrase in each sentence. Write the helping verb(s) on the line.

1. That dog has been barking for two hours. ______________
2. I should finish my homework before dinner. ______________
3. Fred must have eaten something that made him sick. ______________
4. I can run to the corner and back in thirty seconds. ______________
5. Mom is wearing her favorite outfit today. ______________
6. The coach was training the team for next month's race. ______________
7. Dan has been working at the bank for several years. ______________
8. Sarah and David have been planning a surprise party. ______________
9. Usin is driving her car to the amusement park. ______________
10. The teacher could not find my homework assignment. ______________

Sometimes the verb phrase is interrupted by other words. Find the helping verb and underline it once. Underline the main verb twice.

Example: I <u>did</u> eventually <u>find</u> the library.

1. She is never going to finish reading that book.
2. Did you find the piece of pie that I left for you?
3. Have you always lived on Maple Street?
4. Did your cell phone just ring?
5. Has the store opened yet this morning?

Name ______________________ Date ____________

Missing Link

RULE

Verbs don't always show action. These "state-of-being" verbs are forms of the word "be" (be, am, is, are, was, were) and are often used as *linking verbs*. A linking verb connects the subject with another word in the sentence such as another noun or an adjective that describes the subject.

Examples: Fido **is** a beagle.
The lobsters were very expensive.

Other linking verbs include *appear, become, feel, look, smell,* and *seem.*

Example: My computer **appears** to be broken.

Circle the subject and the word that describes it in each sentence. Write the linking verb on the line.

1. That pie smells wonderful! ____________
2. Josie appears to be nervous. ____________
3. Siberian tigers are endangered. ____________
4. The math test seemed very difficult. ____________
5. The boys looked exhausted after their rafting trip. ____________
6. Susanna became ill on the cruise. ____________
7. Lucinda became my best friend this year. ____________
8. Belugas are marine mammals. ____________

Sometimes a verb can be an action verb or a linking verb. Read each pair of sentences and write *action* or *linking* on the line to show how the verb is used.

1. That pizza really tasted good. ____________
   Bob tasted the hot pizza and burned his tongue. ____________
2. The classroom grew silent when the teacher came in. ____________
   My hair grew several inches last year. ____________
3. Sue appeared to be frightened. ____________
   My best friend appeared on my front porch. ____________

Name ______________________________ Date ______________

Past, Present, Future

RULE

In addition to showing action, the tense of a verb tells *when* the action happened. There are three most commonly used tenses.

- *Present tense* shows what the subject is doing right now. For a singular subject (except "I" or "you"), add *s* or *es* to most verbs. For verbs ending in *y*, change the *y* to an *i* and add *es* to form the present tense with a singular subject.

  Examples: They *bake* a cake. He *bakes* a cake. He *carries* the flag.

- *Past tense* shows what the subject did in the past. Form the past tense of most verbs by adding *ed*. If a verb ends with *e*, add *d*. If a verb ends in *y*, change the *y* to *i* and add *ed*. If a verb ends with a vowel and a consonant, double the consonant and add *ed*.

  Examples: They *baked* a cake. He *carried* the flag. I *batted* the ball.

- *Future tense* shows what the subject will do in the future. Use *will* or *shall* with the main verb to form the future tense.

  Example: He *will bake* a cake tomorrow.

Underline the verb in each sentence. Write *past*, *present*, or *future* on each line to show when the action happened.

1. The library closes at 5 p.m. on Friday. ______________
2. My Uncle John visited us last Thanksgiving. ______________
3. Mrs. Garfield will give us a math test on Wednesday. ______________
4. My little brother chased a frog in our backyard. ______________
5. Twyla will be eight years old on her birthday. ______________
6. Tory returned her books to the library. ______________
7. The light bulb will last for several months. ______________
8. A new pet store opened near here last month. ______________

Circle all the verbs in each line that fit the description.

1. **Present** flies learned finds write will lift jogged stretches crossed
2. **Past** moved mixes buried joins will crash settled burns guarded
3. **Future** shall eat hikes will find moved will pour drink shall write sews
4. **Present** ride takes destroyed eats help killed plant forget
5. **Past** protected replaces will hold carried shall move liked lived swim

Name ______________________________ Date ______________

Rule Breakers

In addition to showing action, a verb also tells *when* the action happened. For most verbs, to show that something has already happened *(past tense)*, you add –ed (want → wanted). These are called **regular verbs**. There are some exceptions. If the verb ends in

RULE

- *e*, add *–d* (care → cared)
- *consonant + y*, change the *y* to *i* and add *–ed* (try → tried)
- *a single vowel + a consonant*, double the consonant and add *–ed* (jog → jogged)

Some verbs have special past tense forms. These are called **irregular verbs**.

Examples: bring → brought, come → came, go → went, make → made, run → ran, write → wrote, think → thought, take → took, say → said, win → won

The *past participle* always uses a helping verb such as *have, has,* or *had*. For regular verbs, this is formed by adding *–d* or *–ed* and using a helping verb (lift → have lifted). Irregular verbs do not form the past participle by adding *–d* or *–ed*.

Examples: know → have (had, has) known; steal → have (had, has) stolen; give → have (had, has) given

Supply the *past tense* and *past participle* for each present tense verb.

Present Tense	Past Tense	Past Participle
1 write		
2 break		
3 drive		
4 take		
5 freeze		
6 speak		
7 see		
8 catch		
9 grow		
10 choose		

Write *regular* or *irregular* after each verb.

1. rang ______________
2. stole ______________
3. dressed ______________
4. broke ______________
5. waited ______________
6. knew ______________
7. fell ______________
8. changed ______________
9. stared ______________
10. tore ______________
11. swam ______________
12. talked ______________

Name ______________________________ Date ______________

I Agree

RULE

The subject and verb must agree in number.

- A singular subject takes a singular verb.
- A plural subject takes a plural verb.

Ask yourself: *What is the subject? Is the subject singular or plural?* In the present tense, most verbs add *s* or *es* to form the singular.

Examples (the subject is underlined once and the verb is underlined twice):

- Singular: The ant lives in a large nest.
- Plural: Honeybees fly from flower to flower.

Underline the correct form of the verb in each sentence.

1. Five dogs (roam, roams) through our yard every afternoon.
2. The boy's feet (hurt, hurts) when he wore his brother's shoes.
3. My friends always (squabble, squabbles) when they play kickball.
4. Tim's mom (travel, travels) by bus and train to get to work.
5. Volcanoes often (erupt, erupts) with little notice.
6. Donata (exercise, exercises) before every softball game.
7. My kitten is so scared it (jump, jumps) at every sound it hears.
8. Mary and Elena (sings, sing) in the school chorus.

RULE

Some verbs have special singular and plural forms:

- *be* → singular: is plural: are
- *have* → singular: has plural: have
- *do* → singular: does plural: do

Write the correct form of each verb on the line.

1. **Be** she ______________ they ______________
2. **Have** Juan ______________ jaguars ______________
3. **Do** the puppy ______________ students ______________

Name ______________________________ Date ______________

Pick a Pronoun

RULE

A *pronoun* is a word that takes the place of a noun.

- A *subject pronoun* acts as the subject of the sentence (I, you, he, she, it, we, they).

  Example: He ran two miles yesterday.

- An *object pronoun* acts as the object of the sentence, an indirect object, or the object of a preposition (me, you, him, her, it, us, them). Note: *You* and *it* can be either subject or object pronouns.

  Examples: My sister baked *them.* (object of the sentence)
  Linda gave *him* the book. (indirect object)
  Joe and Lucy are coming with *us.* (object of a preposition)

Circle the pronoun in each sentence. On the line, write *subject* or *object.*

1. Troy helped him rake the leaves in the park. ______________
2. Arianna's teacher asked her a question. ______________
3. They couldn't find the path to the lighthouse. ______________
4. Four girls asked him to the dance. ______________
5. You should be able to finish by seven o'clock. ______________
6. Cali gave them the book yesterday. ______________
7. Mrs. Gonzales sent us an invitation to the party. ______________
8. I ate dinner at a friend's house last night. ______________

Complete each sentence by placing a pronoun in the blank.

1. ______________ didn't eat until after soccer practice.
2. Leslie left ______________ bookbag at the library.
3. ______________ was one of the strangest people Lu had ever seen.
4. Mary and Ana saw ______________ at the mall yesterday.
5. Tomas met ______________ at the practice field after school.
6. Neil looked but was not able to find ______________.

Name ______________________ Date ____________

All Mine

RULE

A *possessive pronoun* shows ownership or possession. Possessive pronouns **do not** require apostrophes.

Examples: *Our* classroom won a prize for collecting the most money.
The red snowboard is *his*.
The dog buried *its* bone under the tree.

Use each possessive pronoun in a sentence.

1. **mine** ______________________
2. **ours** ______________________
3. **your** ______________________
4. **their** ______________________
5. **its** ______________________
6. **hers** ______________________

Review: Circle the correct word in each pair.

1. Alyssa found (her, she) brother hiding in (him, his) treehouse.
2. Skateboarders are required to wear helmets on (they're, their) heads.
3. The kayaker was found at (his, him) campsite after (he, him) capsized.
4. (We, Us) ran three miles every day to practice for (our, your) first race.
5. (He, Him) read the book because he liked the picture on (its, it's) cover.
6. (My, Mine) house is closer to the school than (your, you're) house.
7. (Your, You're) yard is a lot bigger than (ours, ours').
8. Troy sent (my, mine) letter to (his, he's) own address!
9. (They're, Their) first hike was on (their, they're) grandfather's farm.
10. (Our, Ours) photos will be featured in (he's, his) new book.

Name ______________________________ Date ______________

Never Disagree

RULE An *antecedent* is the noun to which a pronoun refers. The pronoun must agree with its antecedent in both number and gender.

Examples: Marco lost his coat on the field trip. The dog tore its bedding into shreds.

Write a pronoun that could refer to each antecedent.

1. Landon and I ______________
2. a car ______________
3. dancers ______________
4. Dan and his brother ______________
5. the raccoon ______________
6. the three boys ______________

Identify the pronoun and antecedent in each sentence.

1. A kitten had its paw stuck in a drainpipe.
2. Aunt Sophie forgot to take her purse to the store.
3. Becky and her friends went camping.
4. Five boys bought their coach a gift.
5. Mr. Marks asked the mechanic to fix his car.
6. Bees help flowers when they collect pollen.
7. Jack and I lost the directions for our new game.

Pronoun	Antecedent

Write a sentence using the antecedent and pronoun provided.

1. (the sun, its) ______________________________
2. (Eli and Alan, their) ______________________________
3. (Larry, his) ______________________________
4. (newspaper, it) ______________________________
5. (my sister, her) ______________________________

Name ______________________________ Date ______________

Describe It

RULE

An *adjective* describes a noun or a pronoun. It tells

- *what kind* → huge, green, fluffy
- *which one* → that, this, those
- *how many* → several, many, ten

Underline the adjective in each sentence. Write the noun it describes in the blank.

1. The path led to the deserted cabin. ______________
2. Nervous warriors got ready for the battle. ______________
3. Luxurious flowers bloomed in the flowerbed. ______________
4. The fidgety toddler jumped out of his seat. ______________
5. Sparkling water gurgled across the rocks in the stream. ______________
6. He climbed the narrow stairway to the attic. ______________
7. Mama couldn't grow flowers in the rocky soil. ______________
8. The uneven terrain caused Marty to trip. ______________

RULE

Sometimes more than one adjective in a row is used to modify or describe a noun.

- Use a comma if the word "and" between the adjectives sounds sensible.

  Example: The *tall, handsome* man played the main character.
- Do not use a comma if the word "and" between the adjectives does not make sense.

  Example: Those huge dogs ran through our yard.

Complete each sentence with two adjectives.

1. ______________ ______________ horses grazed in the field.
2. An ______________, ______________ barn collapsed during the storm.
3. The ______________, ______________ woman sat on the park bench.
4. A ______________, ______________ book sat on the bookshelf.

Name ______________________ Date ____________

It's Only Proper

RULE

A *proper adjective* describes a noun or a pronoun. It is related to a proper noun and must be capitalized.

Example: We visited several *Mexican* villages on our vacation.

Circle the proper adjective in each sentence. Write the noun it describes on the line.

1. The coin she found was Canadian. ____________
2. German dancers starred in the show. ____________
3. The class I am taking is about European history. ____________
4. I wonder if there really is Martian life. ____________
5. British literature is popular worldwide. ____________
6. The Atlantic coastline suffered storms all winter. ____________

Choose five proper adjectives from the box. Use each in a sentence.

Word Box

Alaskan	Irish	Chinese	French
American	Siamese	Congressional	Japanese
Greek	Mexican	Hawaiian	African

1. ______________________________
2. ______________________________
3. ______________________________
4. ______________________________
5. ______________________________

Name ____________________ Date ____________

What's the Difference?

RULE Sometimes a word can be either a *noun* or an *adjective* depending on how it is used in a sentence.

Example: The *gym* was hot and stuffy after the game. (noun)
My *gym* shorts faded in the wash. (adjective)

Circle **A** if the underlined word is an adjective or **N** if it is a noun.

A N 1. My favorite flower is the carnation.

A N 2. The sunflower seeds all sprouted last week.

A N 3. My team captain had to resolve the crisis.

A N 4. Baseball is America's favorite pastime.

A N 5. The subway station was closed for repairs.

A N 6. Mom's glass pitcher fell off the shelf.

A N 7. The tomato vine was over six feet tall.

A N 8. Tom's parents bought him a new football jersey.

Use each word in two sentences. In the first sentence, use the word as a noun, and in the second, use it as an adjective.

1. computer

____

____

2. summer

____

____

3. steel

____

____

Name ____ Date ________________

How Did That Happen?

RULE An *adverb* is a word that describes a verb, an adjective, or another adverb. An adverb answers the questions *how, when*, or *where*. Many adverbs end with *–ly*.

Examples: How? The baby was crying <u>softly</u>.
When? I have to do my homework <u>now</u>.
Where? I'll meet you over <u>there</u>.

Write the adverb that describes the underlined word(s). Does it tell *how, when*, or *where*?

1. My parents <u>should arrive</u> soon.
2. The girls <u>whispered</u> quietly in the library.
3. The neighbors <u>were arguing</u> very loudly.
4. When it rained, we <u>ran</u> inside.
5. I'll <u>study</u> for the math test tomorrow.
6. Jeremy <u>grabbed</u> the dog firmly by the collar.
7. The artist skillfully <u>drew</u> my picture.

Adverb	Question Answered

Complete each sentence with an adverb that answers the question.

1. **How?** The traffic moved ____________________ along the snow-covered highway.
2. **When?** ____________________ the sky turned gray and it began to rain.
3. **How?** The entire class sat ____________________ waiting for the bell to ring.
4. **Where?** Come ____________________ and I'll get you a glass of water.
5. **When?** I will tell you the whole story ____________________.
6. **Where?** He went ____________________ to hide the box of candy.
7. **How?** The snow fell ____________________ on the roof of the cabin.
8. **How?** The new student walked ____________________ to the front of the classroom.

Name ______________________________ Date ______________

More or Less

RULE

Adjectives can be used to compare two or more nouns.

There are two degrees of comparison:

- *Comparative* → most adjectives add *–er* to compare *two* nouns. Sometimes, adjectives with two or more syllables use the modifier *more* or *less*.

  Examples: Marco is taller than Simeon.
  I think football is more exciting than baseball.

- *Superlative* → most adjectives add *–est* to compare *three* or more nouns. Sometimes, adjectives with two or more syllables use the modifier *most* or *least*.

  Examples: Shauna is the tallest girl in her class.
  Maya is the most beautiful person I know.

Underline the comparison adjective or adjective phrase in each sentence. On the line, write *comparative* or *superlative*.

1. The tallest bird that ever lived was New Zealand's giant Moa. ______________
2. One of the most vicious dinosaurs was the velociraptor. ______________
3. Killer bees are more aggressive than honeybees. ______________
4. Wolves are more dangerous when they roam as a pack. ______________
5. The barracuda is one of the most ferocious sea creatures. ______________
6. Green turtles are more plentiful than hawksbill turtles. ______________
7. Some snakes are more poisonous than others. ______________
8. The largest moth has a wingspan of ten inches. ______________

Write the *comparative* and *superlative* forms of each adjective.

Adjective	Comparative	Superlative
1 easy		
2 small		
3 busy		
4 special		
5 terrific		
6 serious		

Name ______________________________ Date ______________

More Often Than Not

RULE Adverbs can also be used to make comparisons.

- Most adverbs add *–er* or *–est* to make comparisons.

  Examples: I finished the test *sooner* than Josie.
  I slept the *latest* of anyone in the group.

- Use *more* or *most* with most adverbs that end in *–ly* or have two or more syllables.

  Examples: Write *more carefully* next time.
  Blue was chosen *most frequently* of all.

- Some adverbs make comparisons in a special way.

  Examples: badly → worse → worst
  well → better → best

Circle the comparison adverb or adverb phrase in each sentence.

1. I studied harder than Terry and I earned a higher grade.
2. You must sit more quietly in church than in a restaurant.
3. Gina ate more quickly than I did.
4. Soccer practice begins earlier than dance class.
5. Mrs. Lipscomb's class worked most carefully of all.
6. The boy who spoke most clearly won the speech contest.
7. I know Jane better than I know her brother Ian.
8. My little brother acts worst when he doesn't take a nap

Write the correct adverb or adverb phrase in the blank.

1. I cleaned my room ____________ (carefully) than my sister.
2. The spotted puppy barks ____________ (loudly) than his littermates.
3. Mrs. Sparks's class behaved ____________ (well) of all in the cafeteria.
4. Trey talked ____________ (often) than anyone in the group.
5. I got to the park ____________ (early) than my friends.

Name ______________________________ Date ______________

Inside and Out

RULE

A *preposition* is a word that shows how a noun or pronoun is related to another word in the sentence. It is a small word that has a big impact on the meaning of the sentence.

Examples: I leaned *against* the wall.
She sat *beside* me.
He walked *across* the street.

These are some of the most common prepositions: *about, above, across, after, along, at, before, behind, below, beneath, beside, between, beyond, by, down, during, for, from, in, inside, near, of, off, on, out, outside, over, through, to, toward, under, up, with, without.*

Prepositions can change the meaning of a sentence. Write two prepositions from the list above that could be used to fill in each blank. The two choices should change the meaning of the sentence.

1. Jody parked her car ___?___ the restaurant. __________ __________
2. I walked ___?___ the church. __________ __________
3. The boys set up a tent ___?___ the trees. __________ __________
4. My puppy hid ___?___ the box. __________ __________
5. Sue called me ___?___ dinner. __________ __________

Find the preposition in each sentence. Write it on the line.

1. Unleashed dogs aren't allowed within city limits. __________
2. Johnny leaned his bicycle against the bike rack. __________
3. I took the package from the delivery man. __________
4. Dad loves cake with his ice cream. __________
5. Trina got sick before gym class yesterday. __________
6. Mike munched popcorn during the movie. __________
7. The colorful fish swam beneath the boat. __________
8. My house is near the local community center. __________
9. The truck was coming straight toward me. __________
10. We all sat around the campfire last night. __________

Name ______________________________ Date ______________

Above and Below

RULE

A *preposition* is always followed by a noun or pronoun called the object of the preposition.

Examples: (the preposition is underlined once and the object of the preposition is underlined twice)
The dog jumped over the fence.
I kicked the ball through the window.

A *prepositional phrase* includes the preposition, the object of the preposition, and all the words in between.

Examples: My brother hid *under the bed.*
I rode my sled *down the hill.*

Underline the prepositional phrase in each sentence. Write the preposition on the line.

1. Jack helped build a bridge across the river. ______________
2. During the Civil War, many families were torn apart. ______________
3. The sun rose over the mountaintops. ______________
4. Mom went to the mall this morning. ______________
5. The flowers wilted because of the high temperatures. ______________
6. I carried my groceries to the car. ______________
7. I ate two packages of raisin cookies. ______________
8. Around the holidays, mom is always very busy. ______________
9. I found my wallet underneath my coat. ______________
10. The hot air balloon soared across the valley. ______________
11. The play was performed by the fourth graders. ______________
12. My kitten was stuck in the tree. ______________
13. The sides of our barn are painted red. ______________
14. The kite's tail was attached with a safety pin. ______________
15. After the ballgame, we all went swimming. ______________

Name ______________________________ Date ______________

What's the Connection?

RULE

A *conjunction* is a word that connects words or groups of words. The most common conjunctions are *and, but,* and *or.*

- A conjunction connects two single words that are the same part of speech.

  Examples: He ate *popcorn* **and** an *apple.* [connects two nouns]
  She **and** *I* read the same book. [connects two pronouns]
  The horse *bucked* **and** *snorted.* [connects two verbs]

- A conjunction connects groups of words.

  Examples: I left my books *at the library* **or** *in Josie's car.*
  I *wrote a letter* **but** *received no reply.*

Circle the conjunction in each sentence. Write the part of speech of the pair of words being connected by each conjunction.

1. Write your name quickly and legibly. ______________
2. The girls built a small but cozy clubhouse. ______________
3. The Indians used mud and grass to make bricks. ______________
4. You may have tacos or tamales for lunch. ______________
5. The car skidded and swerved on the ice. ______________
6. The glass vase fell and broke. ______________
7. He talked loudly and continually. ______________
8. His jaw was red and swollen where he got hit. ______________
9. I stopped the car abruptly but safely behind the wreck. ______________
10. The shed was old but sturdy. ______________

Complete each sentence with the correct word: *and, but,* or *or.*

1. Ladybugs sometimes live in the desert __________ prefer gardens.
2. Dragonflies have short antennae __________ two pairs of wings.
3. Cockroaches can either feed during the daytime __________ at night.
4. Snakeflies eat pollen __________ insects.

Name ______________________________ Date ______________

Yes!

RULE An *interjection* is a word or phrase that shows feeling or emotion. If it shows strong feeling, it is followed by an exclamation point. If it shows mild feeling, it is followed by a comma.

Examples: Wow! This must be the hottest day of the year so far!

Well, you did a nice job on that project.

Circle the interjection in each of these sentences.

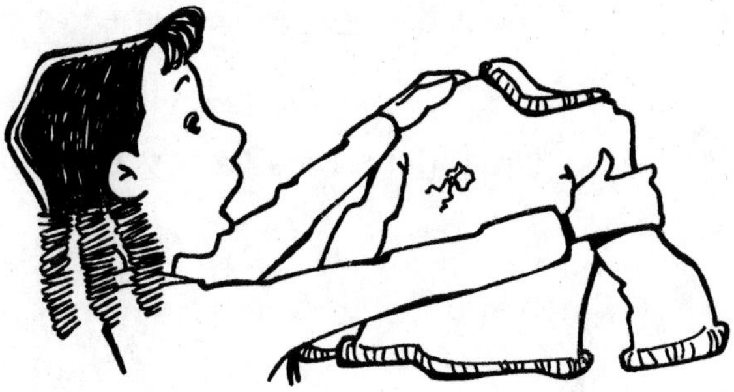

1. Good grief! The wind just blew our tent away!
2. Hurray! Our team finally won two games in a row!
3. Jan cried, "Oh, there's a hole in my favorite sweater."
4. Oops! I just dropped the spoon into the cake batter!
5. Mom cried, "Oh, no! I left my wallet at the grocery store!"
6. Fantastic! I got the highest grade in the whole class!
7. Hey, that was my sandwich you just ate.
8. "Help! I can't hang on much longer!" sobbed the boy on the overturned raft.
9. Darlene screamed, "Ouch! That's my hand you're standing on!"
10. Ugh! Mrs. Smathers is giving us a math test again tomorrow!
11. Wow! Have you ever seen a bear as big as that one?
12. Jan grinned, "Ah, now I see what you mean."
13. Dad yelled, "Surprise! I found the ten-dollar bill you lost last week!"
14. Alas, that was the last of the doughnuts.
15. Amy jumped up and hollered, "Yes! I won first prize in the poetry contest at school!"

Name ______________________________ Date ______________

Practice It!

Directions: Two words are underlined in each sentence. Name the part of speech for each underlined word. Choose from the following: *noun, verb, pronoun, adjective, adverb, preposition, conjunction,* or *interjection.*

1. Julia attended (1) a track meet at the park yesterday (2). ① ____________ ② ____________
2. Wow (1)! That's the scariest (2) costume I've ever seen! ① ____________ ② ____________
3. The salamander (1) crawled across (2) the rock. ① ____________ ② ____________
4. Carmen and (1) Caitlyn found a new (2) box of pencils. ① ____________ ② ____________
5. Usually (1), I don't babysit for my (2) little brother. ① ____________ ② ____________
6. The scrawny (1) puppy climbed onto (2) Mara's lap. ① ____________ ② ____________
7. Ian finished the test (1) more quickly (2) than Sean. ① ____________ ② ____________
8. Haven't you (1) finished eating lunch yet (2)? ① ____________ ② ____________
9. A rainbow appeared (1) over (2) the mountains. ① ____________ ② ____________
10. Jon was fascinated (1) by the beautiful (2) music. ① ____________ ② ____________
11. Can you (1) believe (2) how much that skateboard cost? ① ____________ ② ____________
12. Oops (1), I forgot (2) my homework again. ① ____________ ② ____________
13. Do you want to go shopping with (1) us (2) on Monday? ① ____________ ② ____________
14. I'd love to have a kitten, but (1) I'm allergic to them (2). ① ____________ ② ____________

Name ______________________________ Date ______________

That's Important!

RULE

Capital letters are used to point out important words, such as the pronoun *I* and the first word of a sentence by capitalizing the first letter of the word. Also, use a capital letter for the first letter of proper nouns:

- People, pets, and initials that stand for names (Jody Ellis, Fido, E.M. Harrison).
- Geographical features (streets, cities, states, countries, mountains, parks, islands, bodies of water).
- Days, months, and holidays, but not seasons.
- Things of historical importance, such as battles, wars, events, periods of time, and documents (Ice Age, U.S. Constitution).

Examples: The neighbors and I walked down First Street on Friday.
I know that the Battle of Gettysburg was fought in Pennsylvania during the Civil War.

Underline the words that should be capitalized in each sentence.

1. james and jake smith live in alexandria, virginia.
2. memorial day is always the last monday in may.
3. last friday was sara marshall's birthday.
4. the anderson's camped in yellowstone national park the whole month of june.
5. my dogs fluffy and fifi buried my shoe in the backyard last saturday.
6. erin said that christmas eve falls on a wednesday this year.
7. the original declaration of independence is in washington, d.c.
8. the battle of bunker hill was the bloodiest battle of the revolutionary war.
9. anna jarvis of west virginia campaigned for a national observance of mother's day.
10. terry's birthday is on friday, june 24, this year.
11. we visited england and scotland for two weeks in july.
12. mary's parrot is named polly, and it was a christmas present last december.

Name ______________________________ Date ______________

Is That Proper?

RULE

Here are other words that require capitalization of the first letter of the word:

- Abbreviations for proper nouns (Weds., Dec.)
- Titles of persons and their abbreviations (Mrs. Smith, Dr. Jones, President Lincoln)
- Names of organizations and businesses (Girl Scouts, Wal-Mart)
- Names of planets, stars, and constellations (Mars, Aries)
- Names of buildings, monuments, and memorials (Washington Monument, Eiffel Tower)
- Proper adjectives such as nationalities and religions (Spanish, Catholic)
- Names of awards (Oscar, Nobel Peace Prize)

Circle each letter that should be capitalized.

1. mr. harkins and dr. thomas chaperoned the boy scouts to mt. washington.
2. the leaning tower of pisa is a favorite tourist site for italians.
3. the smithsonian institution is near the jefferson memorial.
4. jerry has a dental appointment with dr. short next wednesday at four o'clock.
5. rosie's cafe is on elm street across from the everclean car wash.
6. somerset high school is next door to the george washington carver museum.
7. senator olafson attended a meeting in the u.s. capitol.
8. both the inuits and aleuts are native to the arctic regions of alaska.
9. southwestern florida is home to dry tortugas national park.
10. st. augustine, the first permanent european settlement in the u.s., was settled by the spanish in 1565.
11. the united nations headquarters is located in new york city.
12. harvard university opened in massachusetts just sixteen years after the pilgrims landed.

Name ______________________________ Date ______________

What's the Title?

RULE The first letter of the first word, last word, and all other important words in the titles of books, magazines, newspapers, short stories, plays, poems, songs, and works of art require capitalization.

Examples: I put an ad in the *Los Angeles Times* last week.
Joe read *Inkheart* and wrote a book report.
Homeward Bound is my favorite movie.
My class sang "God Bless America" in the school play.

Each title contains one error. Circle the error and write the word correctly on the blank.

1. Last summer, I read *Walk two Moons* by Sharon Creech. ______________
2. "The boy in the Box" was the title of my story. ______________
3. Dad sang "Old MacDonald Had a farm" all day in the car. ______________
4. Darla Dimple was a character in *Cats don't Dance.* ______________
5. *The Lion, the witch, and the Wardrobe* is a classic book. ______________
6. We had to memorize *Casey at the bat* by Ernest Thayer. ______________
7. I read about extreme skateboarding in *Sports illustrated.* ______________
8. My class is currently reading *My side of the Mountain.* ______________
9. The winning essay was entitled "How to train Horses." ______________
10. *Ramona's world* is my little sister's favorite book. ______________

Rewrite each of the following titles correctly.

1. *gone with the wind* ______________________________
2. "canoeing on the ohio river" ______________________________
3. *national geographic* ______________________________
4. *dragon rider* ______________________________
5. *san francisco chronicle* ______________________________
6. *honey, i shrunk the kids* ______________________________
7. *the little mermaid* ______________________________

Name ______________________________ Date ______________

A Capital Review

Review: Underline each word that needs to be capitalized in the following sentences.

1. the series of books about harry potter was written by j.k. rowling.
2. senator wilson comes from a small town in georgia.
3. jules verne, a french author, wrote *20,000 leagues under the sea.*
4. the planet jupiter is closer to the sun than uranus.
5. the amazon river runs through brazil, the largest country in south america.
6. both florida and texas border the gulf of mexico.
7. the battle of shiloh was fought in tennessee during the civil war.
8. the sign on dodson's drug emporium said "feb. through apr.—closed for remodeling."
9. the colorado river cuts through grand canyon national park in arizona.
10. st. patrick's cathedral is on fifth avenue in new york city.
11. president john adams was also a signer of the declaration of independence.
12. the statue in the lincoln memorial in washington, d.c., was designed by famed american sculptor daniel chester french.
13. last summer, my mom and i went whitewater rafting on the colorado river.
14. even though i speak french and portuguese, my last name is spanish.
15. mr. and mrs. c. r. timmons just bought a house on second street in reno, nevada.

Circle each letter that should be capitalized in the following phrases.

1	summer baseball season	indian ocean	left on elm street
2	golden gate bridge	speaking german	senator calhoun
3	governor roosevelt	next to wendy's	sun trust bank
4	near the amazon river	works for the fbi	the irish dancers
5	the british flag	the bank of canada	sing "jingle bells"

Name ____ Date ________________

It's a Date!

RULE

A comma is always used

- Between the day of the month and the year. When a date appears in the middle of a sentence, a comma is also placed after the year.

  Example: I met Tori on August 14, 1996, at summer camp.
- Between the name of a city and the state, province, or country. When such place names are used in a sentence, also place a comma after the state, province, or country.

  Example: Tory lived in Kennedy, New York, for eight years.
- After the greeting and closing in a friendly letter.

Add commas where needed in the following letter.

154 Arlington Lane
Leesburg VA 20178

July 14 2007

Dear Danielle

Thank you so much for inviting me to go with you to your aunt's house in Philadelphia Pennsylvania next month. I can't wait to see you and your family again.

I'm a little nervous about riding a train alone. Last time I rode one was on May 25 2000 when my family took me to Boston Massachusetts to see a Red Sox game.

My train is supposed to arrive at 9:00 a.m. on August 14 2007. I hope you will be there to meet me. See you then.

Your friend
Keri Ann

Place commas where needed. Write **C** if the item is correct.

1. _____ March 3 1977
2. _____ Berlin Germany
3. _____ Dear Aunt Sue,
4. _____ Dear Mom
5. _____ May 8, 2001
6. _____ Reston Virginia
7. _____ Ontario Canada
8. _____ Wasco California
9. _____ December 21 1947

Name ______________________________ Date ______________

World Series

RULE

Use a comma

- To separate three or more items in a series.

  Example: My three favorite sports are football, baseball, and soccer.

- To show a pause after an introductory word such as no, yes, or well.

  Example: Yes, my name is Susan Anderson.

- To set off the name of a person being directly addressed.

  Example: John, did you finish mowing the lawn?

Place commas where needed in each sentence.

1. Dad I forgot to feed the dog before we left.
2. The only students with perfect attendance are Malika Seth and Kaitlin.
3. I promise you Mom I'll make my bed every day before school.
4. Well I'm glad that math test is finally behind me.
5. The Phillies Red Sox Blue Jays and Astros all won games last night.
6. Sammy get in here right now!
7. You'd better finish that homework soon Rita.
8. Raul Susana and Joey all came down with chicken pox at the same time.
9. I'm so glad to see you again Stacie because I've really missed you.
10. Look Gordon I found your library book underneath the couch.
11. You'll recognize her because she has blonde hair brown eyes freckles and glasses.
12. No I haven't been to the theater to see the new movie.
13. I think Marcy that you are missing the point.
14. Don't forget to pick up a gallon of milk a roll of stamps and a pepperoni pizza.
15. I've traveled to every state except Alaska Hawaii and Rhode Island.

Name ______________________________ Date ______________

Won't You Join Me?

RULE

Use a comma

- Before a *conjunction* (and, but, or) when it is used to join two simple sentences.

  Example: I designed the doghouse, and Josie built it.
  You can go to the ball game, or you can go on the camping trip.

- To set an *appositive* apart from the rest of the sentence (an *appositive* is a word or phrase that renames the noun or pronoun in front of it).

  Example: Mr. Lewis, our school principal, spoke at the assembly yesterday.
  Hawaii, the only island state, is located in the Pacific Ocean.

Place commas where needed in each sentence. Which rule is being used? Write A on the line for "appositive" or C for "conjunction."

1. Finish your homework and I'll give you a snack. ________
2. Mr. Parson the dog's owner promised to pay for replacing the flowers. ________
3. The bedbug a worldwide pest feeds on humans and domestic animals. ________
4. West Virginia the Mountain State has no natural lakes. ________
5. We went to the seashore with Mary and we had a great time. ________
6. Mario bought a video game and he also bought two comic books. ________
7. I will order pizza and salad but Miranda will probably have a taco. ________
8. Rochester just south of Lake Ontario is a beautiful city. ________
9. My favorite book *The Thief Lord* has just been made into a movie. ________
10. Aunt Judy drove to the mall but she couldn't find a place to park. ________
11. Trey my youngest son has just joined a karate class. ________
12. Jeremy and Brendan the twins who live next door turn five next week. ________
13. Cornelia Funke the author of *Inkspell* was born in Germany. ________
14. The third U.S. president Thomas Jefferson was born in Virginia. ________

Name ______________________________ Date ______________

I'm Possessed!

RULE

An apostrophe is used

- To form the possessive of a noun. For a singular noun, add *'s*. For a plural noun that ends in *s*, just add an apostrophe. For a plural noun that does not end in *s*, add *'s*.

Examples: singular	*boy* + *'s*	→	The *boy's* puppy slept under the chair.
plural	*cars* + *'*	→	The *cars'* windows had all been broken.
	men + *'s*	→	The *men's* names were in alphabetical order.

- To show where one or more letters has been omitted in a contraction.

Examples: did + not = didn't I + will = I'll it + is = it's

Underline the correct form of the possessive in each sentence.

1. The (girl's, girls') clubhouse was their favorite place to meet.
2. The (mice's, mices') tails were long and thin.
3. The amusement (park's, parks') map was hard to understand.
4. Yesterday (afternoon's, afternoons') game went on for hours.
5. (Children's, Childrens') choirs performed all afternoon at the fairgrounds.
6. I hid (Christa's, Christas') bike behind the garage.
7. Margarette took photographs of the (geese's, geeses') nest.
8. The football (player's, players') gear was scattered up and down the sidelines.

Write a contraction for each pair of words shown. On the line, write the letters that have been replaced by the apostrophe.

Words	Contraction	Letters replaced
1 had not		
2 I would		
3 you are		
4 she will		
5 we are		
6 I am		
7 he is		

Name ______________________________ Date ______________

You Said It!

RULE

A *direct quotation* is the exact words that a person says and should be written in a certain way:

- Use quotation marks before and after a person's exact words.
- Capitalize the first letter of the first word of each sentence of a quotation.
- Use a comma to separate the quotation from the rest of the sentence (inside the closing quotation marks).
- Place the ending punctuation inside the closing quotation marks if the quote is at the end of the sentence.

Examples: "I finished my English homework," said Amy, "and now I'm tired."
Heidi asked, "Do you think we'll ever finish this job?"
"That's not fair!" shouted Maria. "I didn't get a piece of cake."

Rewrite each quotation with correct punctuation and capitalization.

1. the teacher said don't forget to take your math book home tonight

2. did you know that tomorrow is my birthday asked Sam

3. the track team hasn't practiced this week complained the coach because of the rain

4. that was the best book I have ever read exclaimed Ruth

5. Jonah asked why didn't you help me finish mowing the lawn

6. after the test said the teacher you may read your library books

Name ______________________________ Date ______________

Signals

RULE

Use a colon

- To separate the hour and minute.

  Example: Your appointment is at *3:15* p.m.

- To introduce a list of items at the end of a sentence. Words such as *the following, as follows,* or *these* are used to introduce the list.

  Example: For the hike, you'll need the following: canteen, raincoat, sunscreen, water, and snacks.

Use a semicolon

- To join the parts of a compound sentence when no conjunction is used.

  Example: Don likes skateboarding; I prefer mountain biking.

Place a colon or semicolon where needed in the following sentences.

1. My baby brother was born at 1215 a.m. on January 8, 2005.
2. My favorite seafood is shrimp Mary's is lobster.
3. To make grandma's cookie recipe you'll need to buy the following chocolate chips, raisins, butter, flour, and brown sugar.
4. I love animals horses are my favorite.
5. Summer camp offers activities such as these mountain climbing, wakeboarding, snorkeling, and whitewater rafting.
6. Soccer practice lasts from 525 until 730 tonight.
7. Alaska is the largest state Rhode Island is the smallest.
8. Get your handlebars straightened otherwise, you might have a wreck.
9. Everyone from my class came to the picnic except the following students Mary Ann, Juanita, Carly, and Simon.
10. Jerry works in the record store all day at night, he takes college classes.
11. The workshop started at 1000 a.m. and ended at noon.
12. The shirt fits fine the pants are a little too short.

Name ______________________________ Date ______________

One-Way Street

RULE

The *hyphen* is used

- At the end of a line to divide a word that won't all fit on the line. Always divide the word between syllables.
- In compound numbers from twenty-one to ninety-nine.
- In fractions and cardinal numbers that are two words long, such as two-thirds and twenty-first.
- In certain compound words, such as mother-in-law and forget-me-not. You may need to check a dictionary.

Add a hyphen where needed in the underlined words. If the sentence is correct, write C on the line. Use a dictionary if you need to check a word.

1. The car shakes if it goes above sixty five miles per hour. ________
2. My brother in-law's name is Ben Taylor. ________
3. Seventy-five percent of students passed the final exam in history class. ________
4. First class tickets cost more than business class on most airlines. ________
5. Always use an up-to-date encyclopedia when writing a science report. ________
6. I have between eighty and eighty five baseball cards in my collection. ________
7. My dad's first part time job was as a newspaper delivery boy. ________
8. Harry Truman was the thirty-third president of the United States. ________
9. Three fourths of the kids in my class came down with the flu last week. ________
10. My little sister loves to ride on the merry go round. ________
11. I found sixty-seven cents in the bottom of my dresser drawer. ________
12. The best trip that I ever took was last year when my family took me to the astron omy exhibit at the Smithsonian Air and Space Museum. ________

Name ______________________________ Date ______________

Putting It Together

Directions: Rewrite each sentence. Correct all mistakes.

1. kate and me are going to dr. shorts office monday

____

2. of all six boys joseph learned spanish quicker

____

3. mrs allens baby was born on march 12 2007

____

4. the new kittens was born sometime in august

____

5. that american elm tree has grew almost three feet

____

6. keri seen a car wreck on johnson avenue tuesday

____

7. elise watch the movie the lion king three times

____

8. two weeks at summer camp was the title of elis story

____

9. tyler read henry and ribsy by beverly cleary

____

10. pizza is fine sighed arnie but i was hoping for spaghetti

____

Answer Key

Sentence or Not? (page 4)

1. S	**5.** F	**9.** F
2. F	**6.** S	**10.** S
3. F	**7.** F	**11.** F
4. S	**8.** F	**12.** F

1–5 Answers will vary.

Sentence Sense (page 5)

1. (.) declarative
2. (.) declarative
3. (?) interrogative
4. (.) imperative
5. (?) interrogative
6. (?) interrogative
7. (.) declarative
8. (.) imperative
9. (!) exclamatory
10. (?) interrogative
11. (.) declarative
12. (.) imperative

Write It Down! (page 6)

Answers will vary.

The Subject Is (page 7)

1. The timber rattlesnake
   Simple: rattlesnake
2. Meat-eating animals
   Simple: animals
3. The fluffy young owl
   Simple: owl
4. Flat tropical grasslands
   Simple: grasslands
5. The unusual coconut crab
   Simple: crab
6. The brand new Metro Zoo
   Simple: Metro Zoo
7. A fully grown Florida manatee
   Simple: manatee
8. The high misty mountains of central China
   Simple: mountains
9. One of the scariest nighttime sounds
   Simple: One
10. The tallest water bird on the planet
    Simple: bird
11. The white-tailed deer fawn
    Simple: fawn
12. Huge herds of grazing buffalo
    Simple: herds

Predicate Predicament (page 8)

Sentences will vary. Simple predicates are listed below.

1. plays	**6.** resembled
2. was	**7.** built
3. writes	**8.** looked
4. fell	**9.** chewed
5. ate	**10.** fit

What a Combination (page 9)

Answers may vary slightly.

1. A coyote is a member of the dog family, and it is less than half the size of a wolf.
2. An eel looks like a snake, but it is really a type of fish.
3. An Asian elephant is ten feet tall, but an African elephant is even bigger.
4. A hermit crab is a crustacean, and it has five pairs of jointed legs.
5. The female hyena is bigger than the male, and she leads a group called a clan.
6. Seals are quite clumsy on land, but they are fast and graceful in the water.
7. A male ostrich stands eight feet tall, and it can weigh more than 400 pounds.
8. Grasshoppers have wings and can fly, but they usually walk and hop.
9. Dolphins are toothed whales, and there are over thirty different species.
10. Arabian camels have one hump, but Asian camels have two humps.

Identifying Nouns (page 10)

Answers will vary.

How Many? (page 11)

1. wharves, banjos, spies, rodeos
2. hat, hero, studio, echo, piano
3. mosquito, salad, ballerina
4. perches, ladies, flies

Review:

1. women, cities, movies, geese, leaves
2. Darla, Neptune, Garfield, Seattle, Brazil
3. Toyotas, Canadians, Mondays, Buddhists, Yankees
4. pencil, kangaroo, seamstress, equator, frontier

More Than One (page 12)

1. wolves	**1.** ox
2. women	**2.** cactus
3. shelves	**3.** loaf
4. guys	**4.** berry
5. feet	**5.** tomato
6. elves	**6.** calf
7. families	**7.** life
8. monkeys	**8.** moose
9. journeys	**9.** salmon
10. matches	**10.** fly
11. foxes	**11.** fairy
12. churches	**12.** die

Lights, Camera, Action! (page 13)

Underlined verbs are listed below. Alternatives suggested by students will vary.

1. gathered
2. practiced
3. parked
4. assigned
5. waded
6. munched
7. found
8. read
9. rebuilt
10. broke
11. decorated
12. washed

Help! (page 14)

1. has been barking (has been)
2. should finish (should)
3. must have eaten (must have)
4. can run (can)
5. is wearing (is)
6. was training (was)
7. has been working (has been)
8. have been planning (have been)
9. is driving (is)
10. could find (could)

1. is going finish
2. Did find
3. Have lived
4. Did ring
5. Has opened

Missing Link (page 15)

1. pie—wonderful → smells
2. Josie—nervous → appears
3. tigers—endangered → are
4. test—difficult → seemed
5. boys—exhausted → looked
6. Susanna—ill → became
7. Lucinda—friend → became
8. Belugas—mammals → are

1. linking, action
2. linking, action
3. linking, action

Past, Present, Future (page 16)

1. closes → present
2. visited → past
3. will give → future
4. chased → past
5. will be → future
6. returned → past
7. will last → future
8. opened → past

1. flies, finds, write, stretches
2. moved, buried, settled, guarded
3. shall eat, will find, will pour, shall write
4. ride, takes, eats, help, plant, forget
5. protected, carried, liked, lived

Rule Breakers (page 17)

1. wrote, (has, have, had) written
2. broke, (has, have, had) broken
3. drove, (has, have, had) driven
4. took, (has, have, had) taken
5. froze, (has, have, had) frozen
6. spoke, (has, have, had) spoken
7. saw, (has, have, had) seen
8. caught, (has, have, had) caught
9. grew, (has, have, had) grown
10. chose, (has, have, had) chosen

1. irregular
2. irregular
3. regular
4. irregular
5. regular
6. irregular
7. irregular
8. regular
9. regular
10. irregular
11. irregular
12. regular

I Agree (page 18)

1. roam
2. hurt
3. squabble
4. travels
5. erupt
6. exercises
7. jumps
8. sing

1. she is, they are
2. Juan has, jaguars have
3. the puppy does, students do

Pick a Pronoun (page 19)

1. him (object)
2. her (object)
3. They (subject)
4. him (object)
5. You (subject)
6. them (object)
7. us (object)
8. I (subject)

1–6 Answers will vary.

All Mine (page 20)

1–6 Answers will vary.

1. her, his
2. their
3. his, he
4. We, our
5. He, its
6. My, your
7. Your, ours
8. my, his
9. Their, their
10. Our, his

Never Disagree (page 21)

Answers may vary:

1. we, our
2. it, its
3. they, them, their
4. they, them, their
5. it, its
6. they, them, their

1. its—kitten
2. her—Aunt Sophie
3. her—Becky
4. their—boys
5. his—Mr. Marks
6. they—Bees
7. our—Jack and I

1–5 Answers will vary.

Describe It (page 22)

1. <u>deserted</u> cabin
2. <u>Nervous</u> warriors
3. <u>Luxurious</u> flowers
4. <u>fidgety</u> toddler
5. <u>Sparkling</u> water
6. <u>narrow</u> stairway
7. <u>rocky</u> soil
8. <u>uneven</u> terrain

1–4 Answers will vary.

It's Only Proper (page 23)

1. Canadian coin
2. German dancers
3. European history
4. Martian life
5. British literature
6. Atlantic coastline

1–5 Answers will vary.

What's the Difference? (page 24)

1. N
2. A
3. A
4. N
5. A
6. A
7. A
8. A

Sentences: Answers will vary.

How Did That Happen? (page 25)

1. soon—when
2. quietly—how
3. loudly—how
4. inside—where
5. tomorrow—when
6. firmly—how
7. skillfully—how

1-8 Answers will vary.

More or Less (page 26)

1. tallest (superlative)
2. most vicious (superlative)
3. more aggressive (comparative)
4. more dangerous (comparative)
5. most ferocious (superlative)
6. more plentiful (comparative)
7. more poisonous (comparative)
8. largest (superlative)

1. easier, easiest
2. smaller, smallest
3. busier, busiest
4. more special, most special
5. more terrific, most terrific
6. more serious, most serious

More Often Than Not (page 27)

1. harder
2. more quietly
3. more quickly
4. earlier
5. most carefully
6. most clearly
7. better
8. worst

1. more carefully
2. more loudly
3. best
4. more often
5. earlier

Inside and Out (page 28)

1-5 Answers will vary.

1. within
2. against
3. from
4. with
5. before
6. during
7. beneath
8. near
9. toward
10. around

Above and Below (page 29)

1. across the river, across
2. During the Civil War, During
3. over the mountaintops, over
4. to the mall, to
5. of the high temperatures, of
6. to the car, to
7. of raisin cookies, of
8. Around the holidays, Around
9. underneath my coat, underneath
10. across the valley, across
11. by the fourth graders, by
12. in the tree, in
13. of our barn, of
14. with a safety pin, with
15. After the ballgame, After

What's the Connection? (page 30)

1. and (adverbs)
2. but (adjectives)
3. and (nouns)
4. or (nouns)
5. and (verbs)
6. and (verbs)
7. and (adverbs)
8. and (adjectives)
9. but (adverbs)
10. but (adjectives)

1. but
2. and
3. or
4. and

Yes! (page 31)

1. Good grief
2. Hurray
3. Oh
4. Oops
5. Oh, no
6. Fantastic
7. Hey
8. Help
9. Ouch
10. Ugh
11. Wow
12. Ah
13. Surprise
14. Alas
15. Yes

Practice It! (page 32)

1. verb, adverb
2. interjection, adjective
3. noun, preposition
4. conjunction, adjective
5. adverb, pronoun
6. adjective, preposition
7. noun, adverb
8. pronoun, adverb
9. verb, preposition
10. verb, adjective
11. pronoun, verb
12. interjection, verb
13. preposition, pronoun
14. conjunction, pronoun

That's Important! (page 33)

1. James, Jake, Smith, Alexandria, Virginia
2. Memorial, Day, Monday, May
3. Last, Friday, Sara, Marshall's
4. The, Anderson's, Yellowstone, National, Park, June
5. My, Fluffy, Fifi, Saturday
6. Erin, Christmas, Eve, Wednesday
7. The, Declaration, Independence, Washington, D.C.
8. The, Battle, Bunker, Hill, Revolutionary, War
9. Anna, Jarvis, West, Virginia, Mother's, Day
10. Terry's, Friday, June
11. We, England, Scotland, July
12. Mary's, Polly, Christmas, December

Is That Proper? (page 34)

1. Mr., Harkins, Dr., Thomas, Boy, Scouts, Mt., Washington
2. The, Leaning, Tower, Pisa, Italians
3. The, Smithsonian, Institution, Jefferson, Memorial
4. Jerry, Dr., Short, Wednesday
5. Rosie's, Café, Elm, Street, Everclean, Car, Wash
6. Somerset, High, School, George, Washington, Carver, Museum
7. Senator, Olafson, U.S., Capitol
8. Both, Inuits, Aleuts, Arctic, Alaska
9. Southwestern, Florida, Dry, Tortugas, National, Park
10. St., Augustine, European, U.S., Spanish
11. The, United, Nations, New, York, City
12. Harvard, University, Massachusetts, Pilgrims

What's the Title? (page 35)

1. Two
2. Boy
3. Farm
4. Don't
5. Witch
6. Bat
7. Illustrated
8. Side
9. Train
10. World

1. *Gone with the Wind*
2. "Canoeing on the Ohio River"
3. *National Geographic*
4. *Dragon Rider*
5. *San Francisco Chronicle*
6. *Honey, I Shrunk the Kids*
7. *The Little Mermaid*

A Capital Review (page 36)

1. The, Harry, Potter, J.K., Rowling
2. Senator, Wilson, Georgia
3. Jules, Verne, French, Leagues, Under, Sea
4. The, Jupiter, Uranus
5. The, Amazon, River, Brazil, South, America
6. Both, Florida, Texas, Gulf, Mexico
7. The, Battle, Shiloh, Tennessee, Civil, War
8. The, Dodson's, Drug, Emporium, Feb., Apr.,
9. The, Colorado, River, Grand, Canyon, National, Park, Arizona
10. St., Patrick's, Cathedral, Fifth, Avenue, New, York, City
11. President, John, Adams, Declaration, Independence
12. The, Lincoln, Memorial, Washington, D.C., American, Daniel, Chester, French
13. Last, I, Colorado, River
14. Even, I, French, Portuguese, Spanish
15. Mr., Mrs., C.R., Timmons, Second, Street, Reno, Nevada

1. Indian, Ocean, Elm, Street
2. Golden, Gate, Bridge, German, Senator, Calhoun
3. Governor, Roosevelt, Wendy's, Sun, Trust, Bank
4. Amazon, River, FBI, Irish
5. British, Bank, Canada, Jingle, Bells

It's a Date! (page 37)

Add commas after each of the following:

1. Leesburg
2. 14
3. Danielle
4. Philadelphia
5. Pennsylvania
6. 25
7. 2000
8. Boston
9. Massachusetts
10. 14
11. friend

1. 3,	6. Reston,
2. Berlin,	7. Ontario,
3. C	8. Wasco,
4. Mom,	9. 21,
5. C	

World Series (page 38)

1. Dad,
2. Malika, Seth,
3. you, Mom,
4. Well,
5. Phillies, Red Sox, Blue Jays,
6. Sammy,
7. soon,
8. Raul, Susana,
9. again, Stacie,
10. Look, Gordon,
11. hair, eyes, freckles,
12. No,
13. think, Marcy,
14. milk, stamps,
15. Alaska, Hawaii,

Won't You Join Me? (page 39)

1. C—homework,
2. A—Parson, owner,
3. A—bedbug, pest,
4. A—Virginia, State,
5. C—Mary,
6. C—game,
7. C—salad,
8. A—Rochester, Ontario,
9. A—book, *Lord*,
10. C—mall,
11. A—Trey, son,
12. A—Brendan, door,
13. A—Funke, *Inkspell*,
14. A—president, Jefferson,

I'm Possessed! (page 40)

1. girls'	1. hadn't—o
2. mice's	2. I'd—woul
3. park's	3. you're—a
4. afternoon's	4. she'll—wi
5. Children's	5. we're—a
6. Christa's	6. I'm—a
7. geese's	7. he's—i
8. players'	

You Said It! (page 41)

1. The teacher said, "Don't forget to take your math book home tonight."
2. "Did you know that tomorrow is my birthday?" asked Sam.
3. "The track team hasn't practiced this week," complained the coach, "because of the rain."
4. "That was the best book I have ever read!" exclaimed Ruth.
5. Jonah asked, "Why didn't you help me finish mowing the lawn?"
6. "After the test," said the teacher, "you may read your library books."

Signals (page 42)

1. 12:15	7. state;
2. shrimp;	8. straightened;
3. following:	9. students:
4. animals;	10. day;
5. these:	11. 10:00
6. 5:25 7:30	12. fine;

One-Way Street (page 43)

1. sixty-five
2. brother-in
3. C
4. First-class
5. C
6. eighty-five
7. part-time
8. thirty-third
9. Three-fourths
10. merry-go-round
11. C
12. astron-omy

Putting It Together (page 44)

1. Kate and I are going to Dr. Short's office Monday.
2. Of all six boys, Joseph learned Spanish quickest.
3. Mrs. Allen's baby was born on March 12, 2007.
4. The new kittens were born sometime in August.
5. That American elm tree has grown almost three feet.
6. Keri saw a car wreck on Johnson Avenue Tuesday.
7. Elise watched the movie *The Lion King* three times.
8. "Two Weeks at Summer Camp" was the title of Eli's story.
9. Tyler read *Henry and Ribsy* by Beverly Cleary.
10. "Pizza is fine," sighed Arnie, "but I was hoping for spaghetti."